Leafy Notes

A Few Notes about Plant Leaves

A Coloring Book

By Michael Reed

Leafy Notes

A Few Notes about Plant Leaves

A Coloring Book

By Michael Reed

Copyright ©2018 by MR

All Rights Reserved.

Acknowledgment

I thank God for giving me the interest on this subject.

A leaf is an important plant organ for us and Earth.

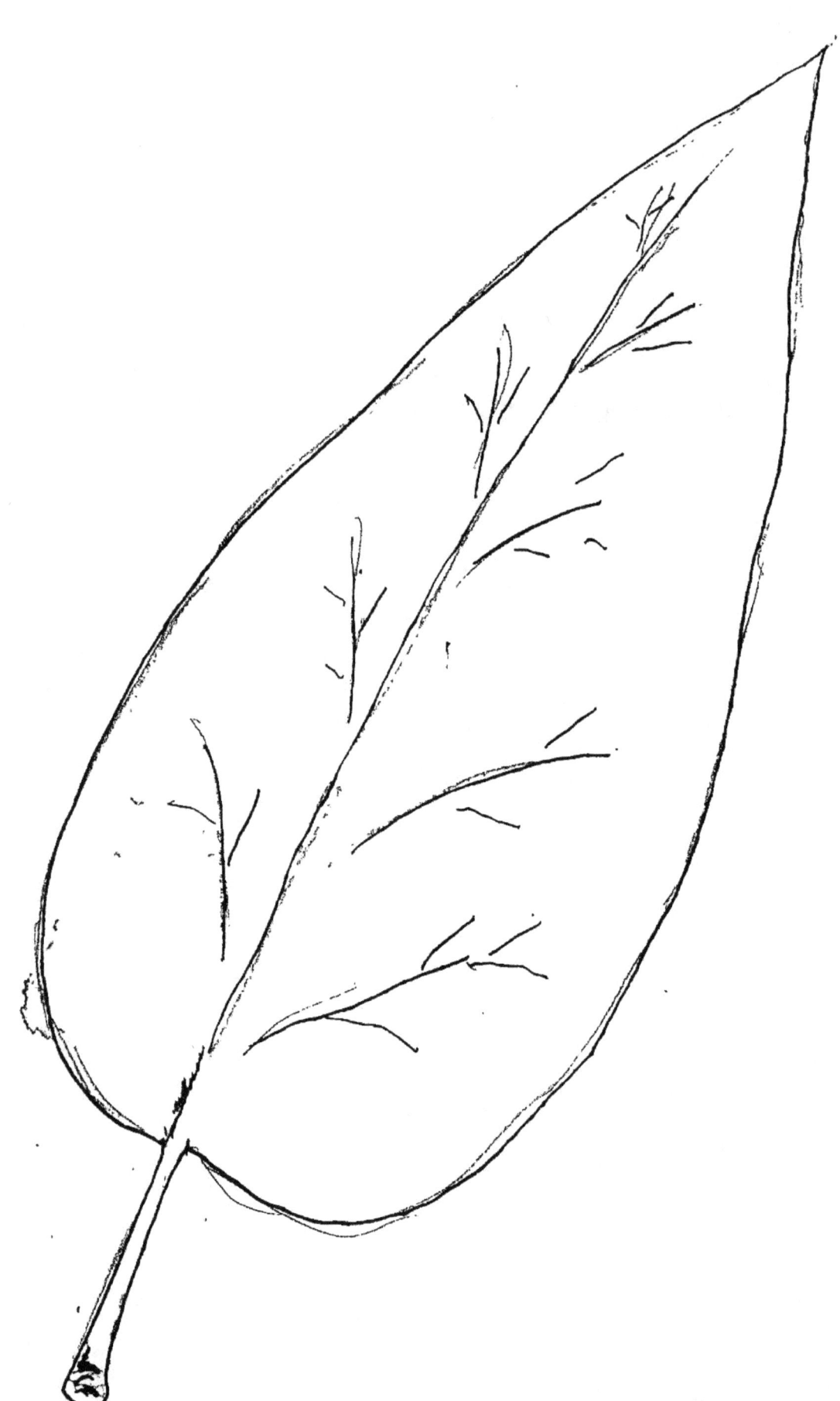

Each plant has its own leaf kind.

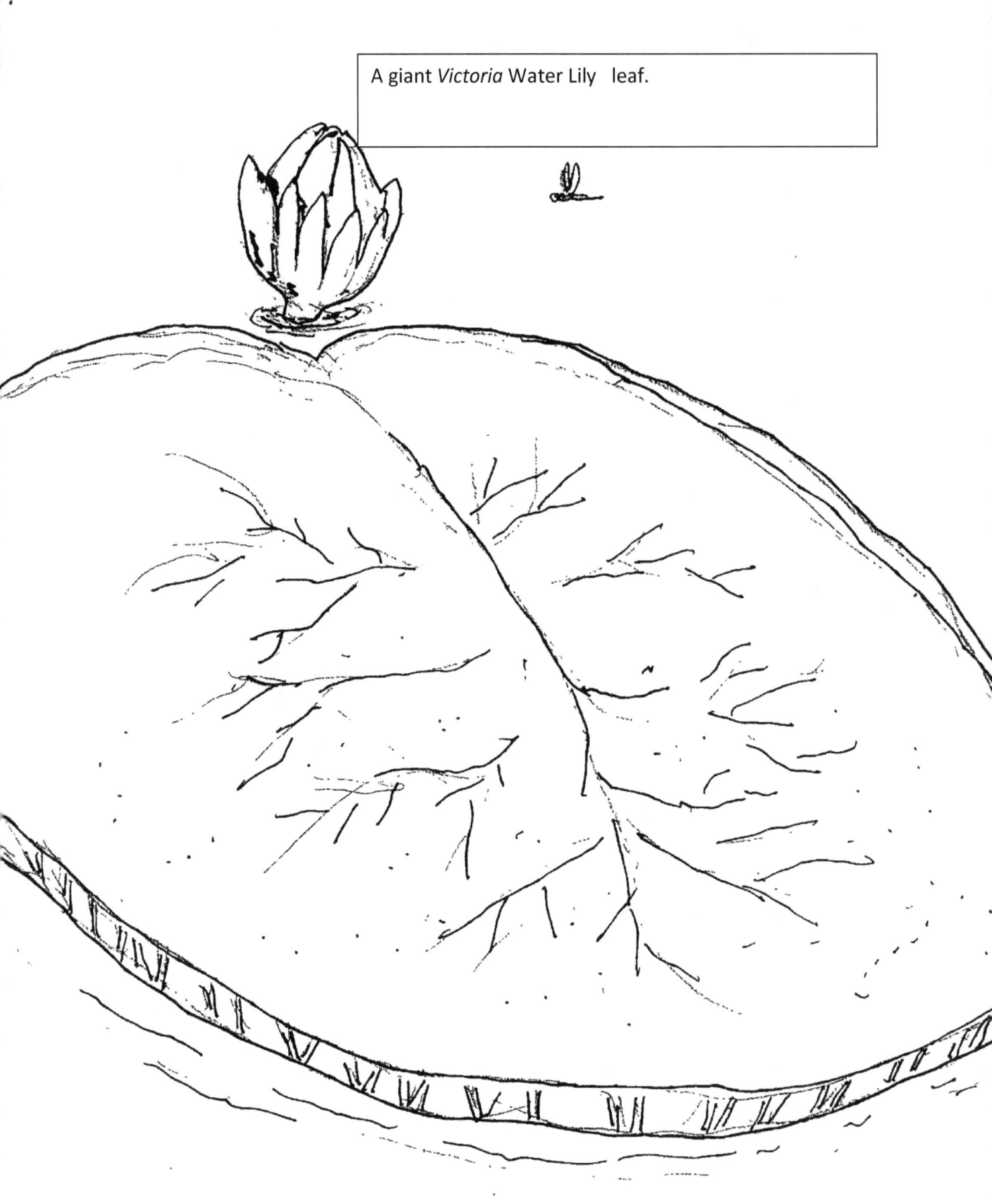

A giant *Victoria* Water Lily leaf.

Venus Fly trap (*Dionaea muscipula*). This plant was made to trap insects for its survival.

Like the Venus Fly Trap, a Pitcher Plant has leaves that are designed to attract and trap insects and other small animals for its nutrients.

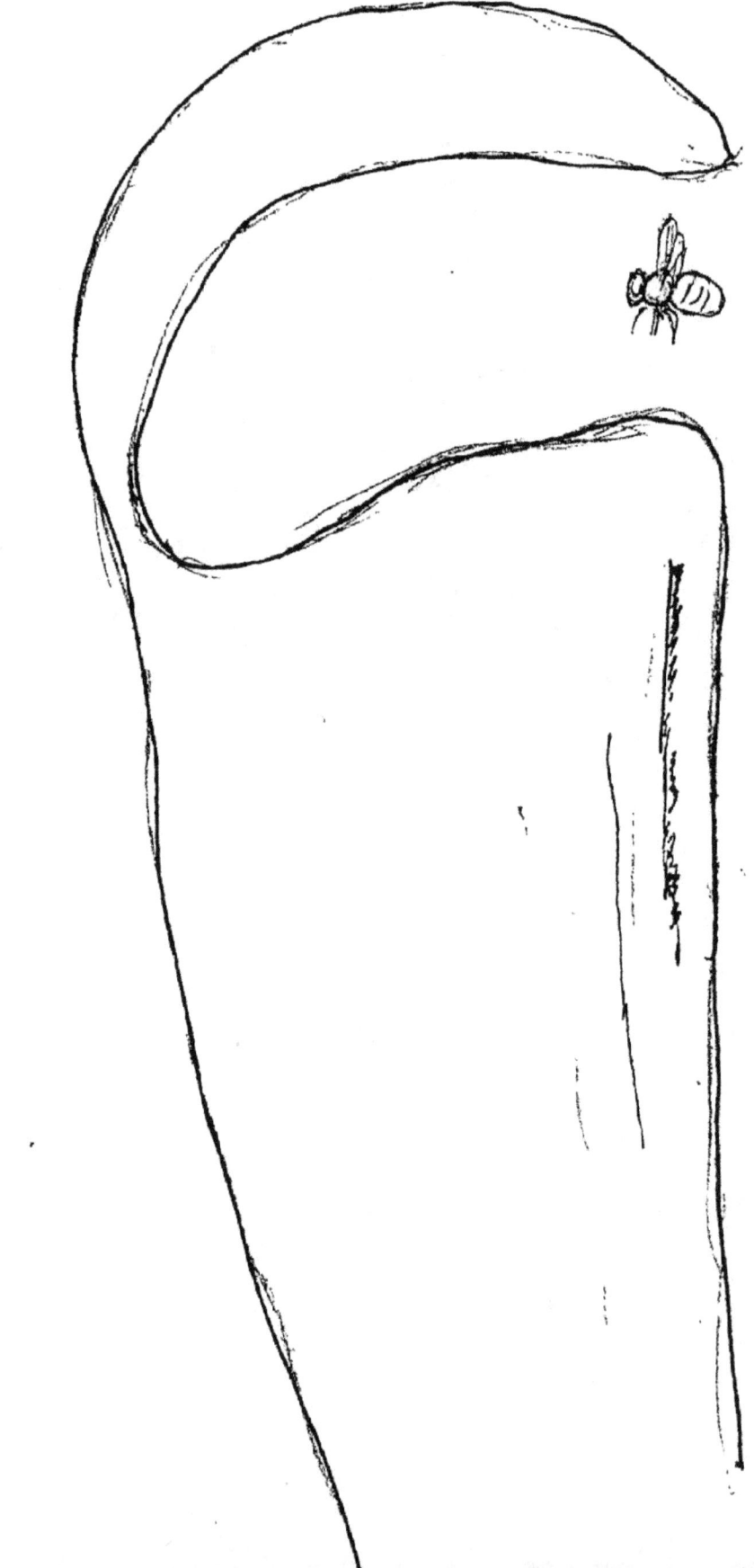

Ginkgo biloba leaf.

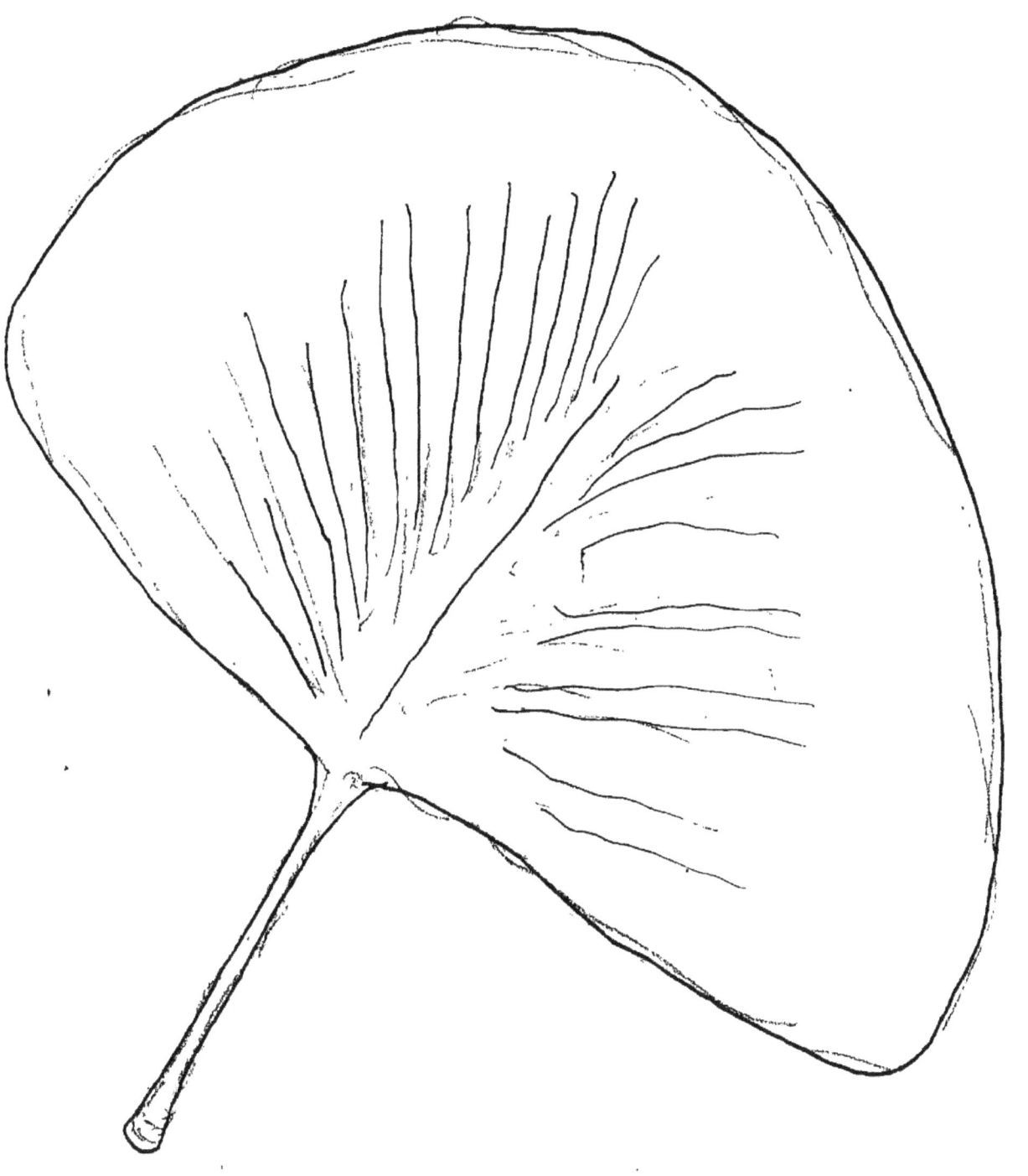

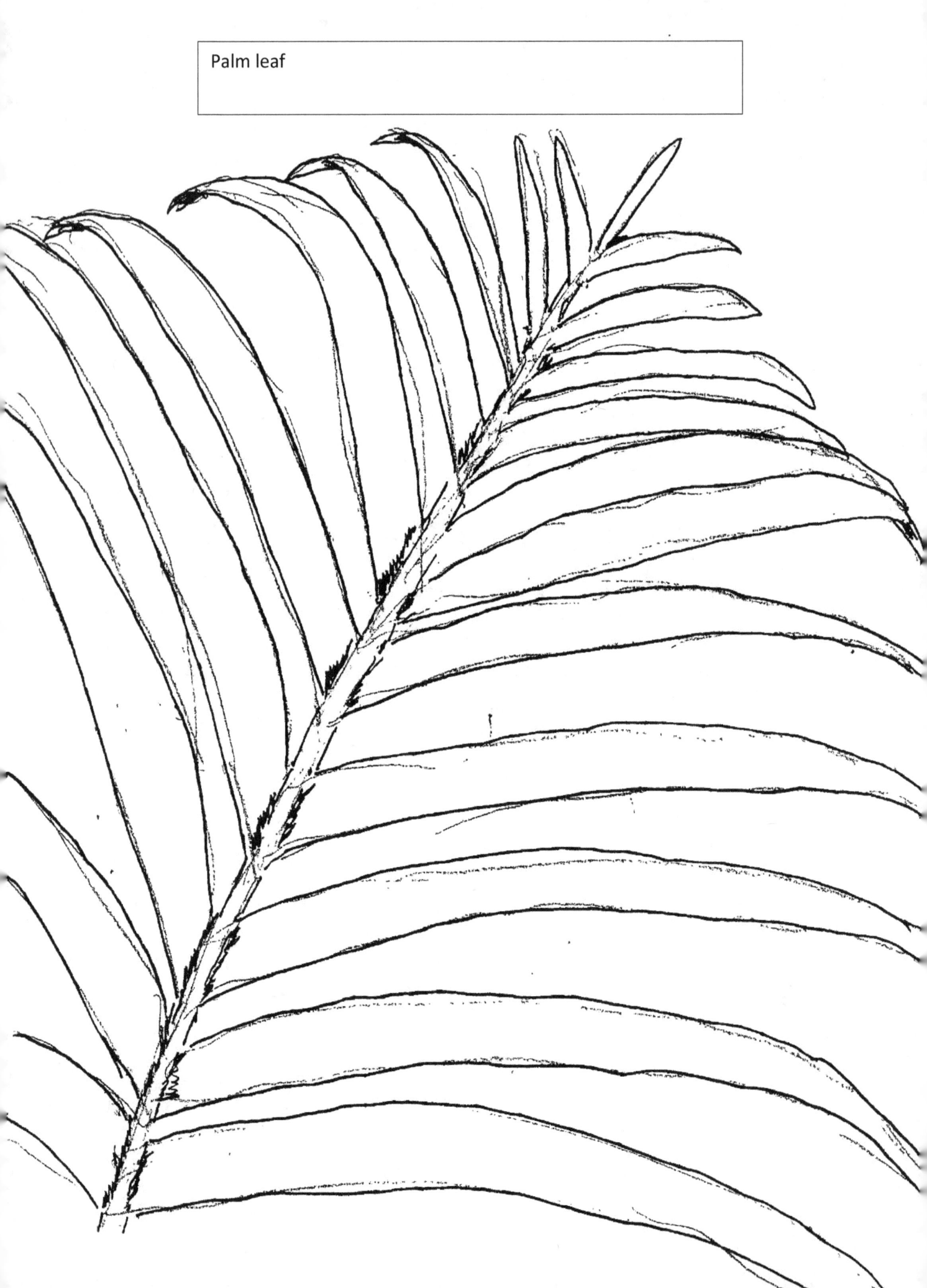
Palm leaf

Oak Leaf

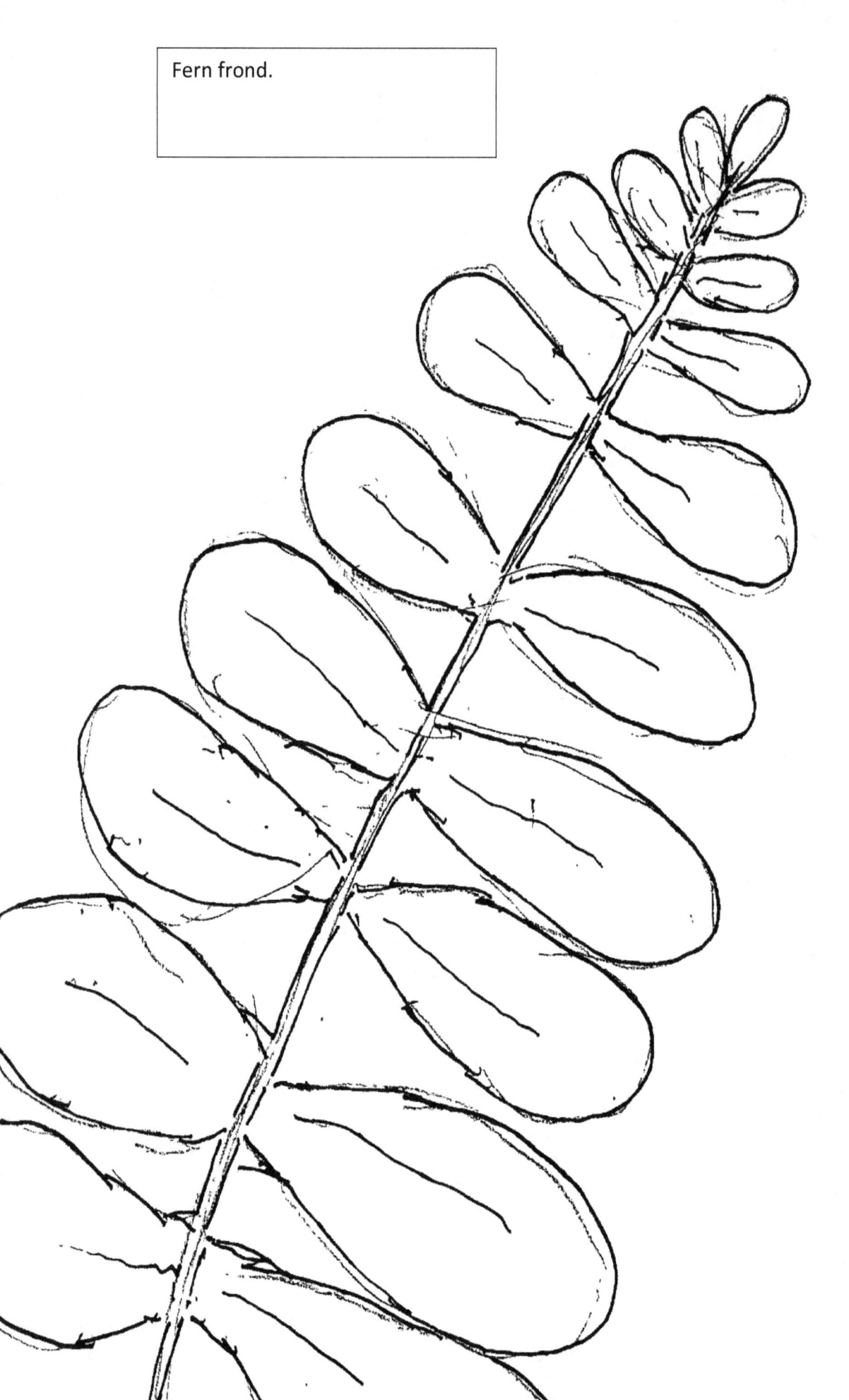

Fern frond.

Many plants use leaves to turn water, carbon dioxide, and sunlight into their food and oxygen for the Earth.

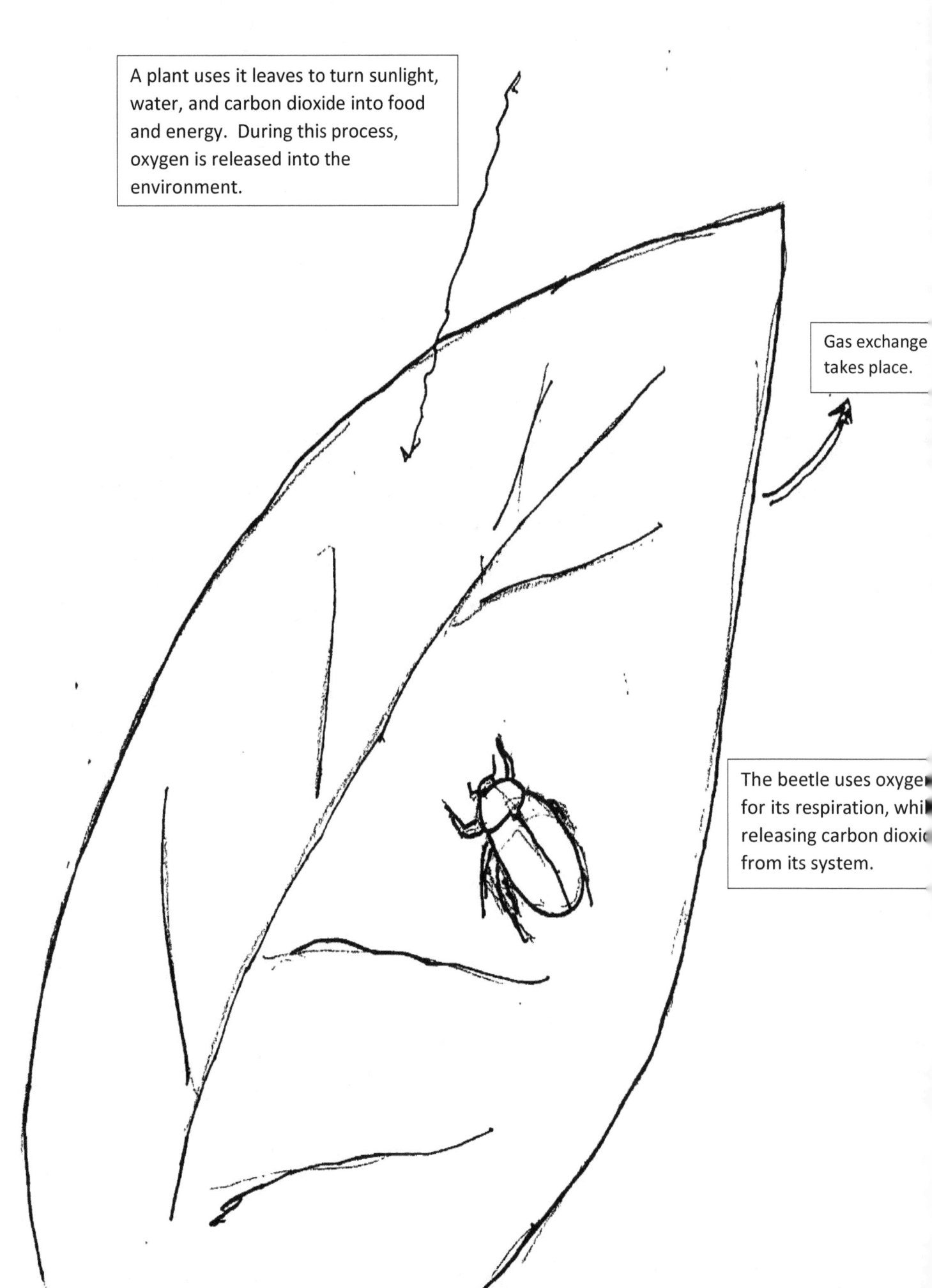

We use some plant leaves for ...

...food

A can and bowl of spinach.

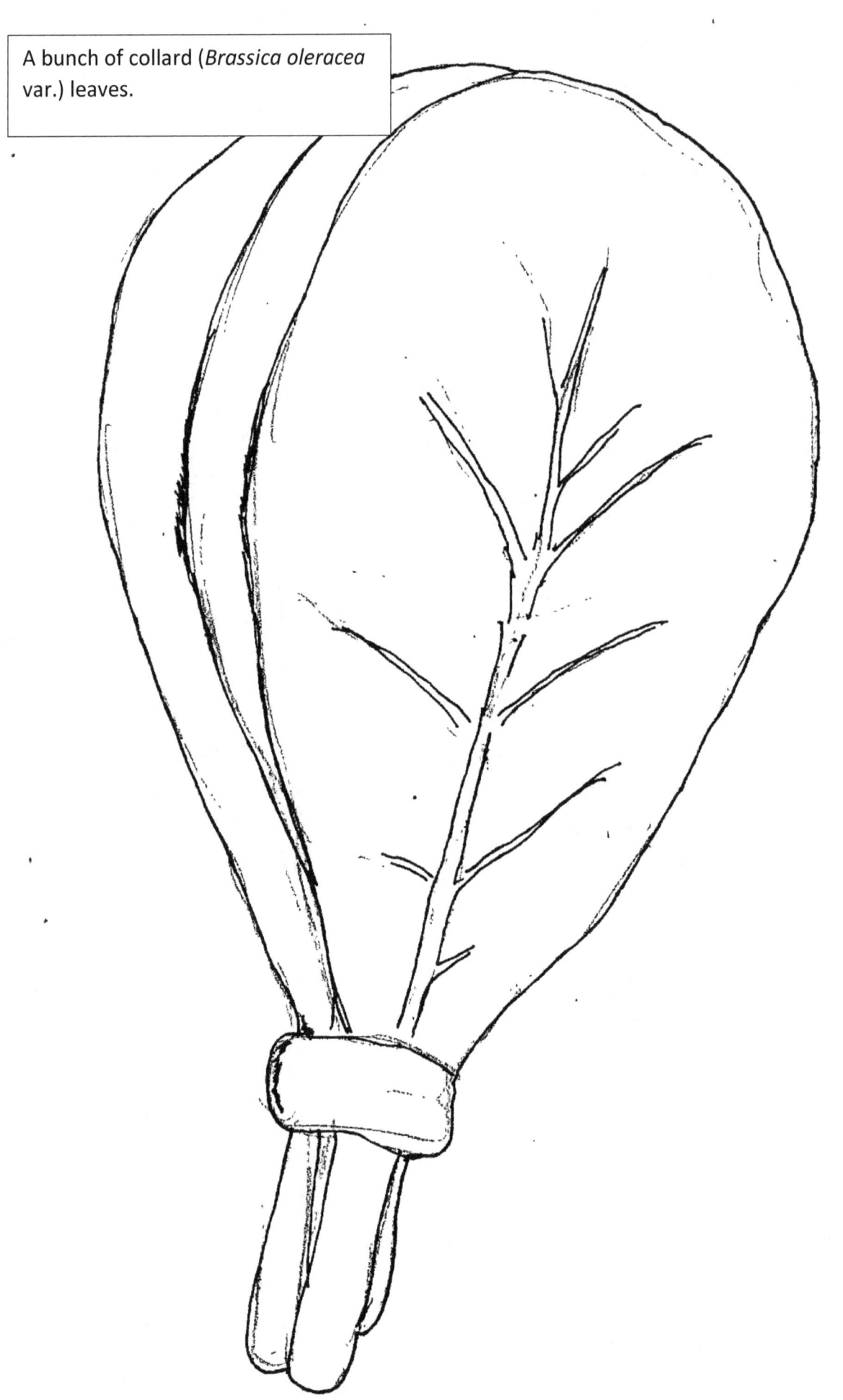

A bunch of collard (*Brassica oleracea* var.) leaves.

A head of cabbage (*Brassica oleracea* var.).

Lettuce leaves in a salad bowl.

...medicine

An *Aloe vera* leaf. This plant is used to make materials for treating wounds and health drinks.

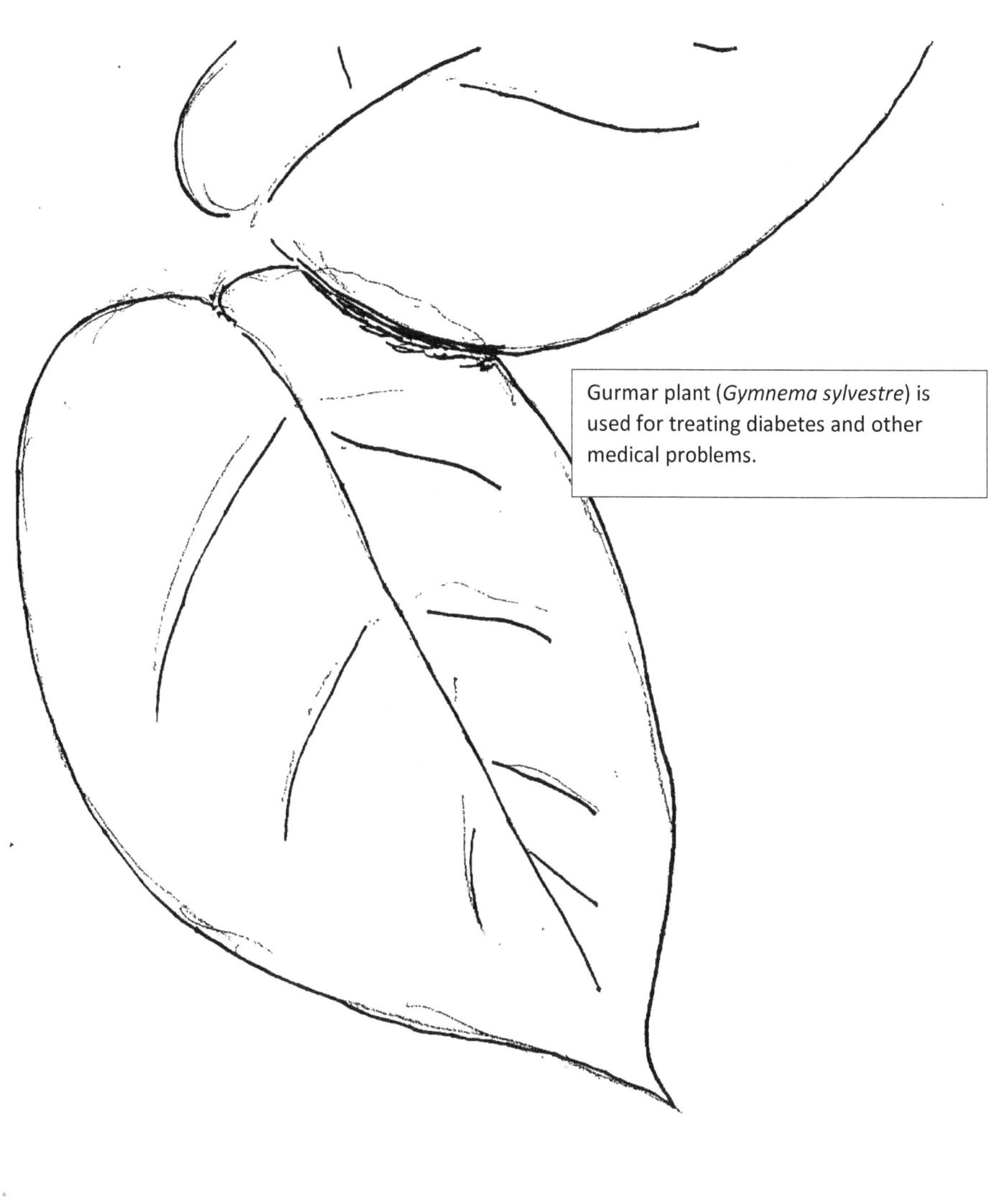

Gurmar plant (*Gymnema sylvestre*) is used for treating diabetes and other medical problems.

...and shelter.

In the tropics, some people used palm leaves as roofing materials for their houses.

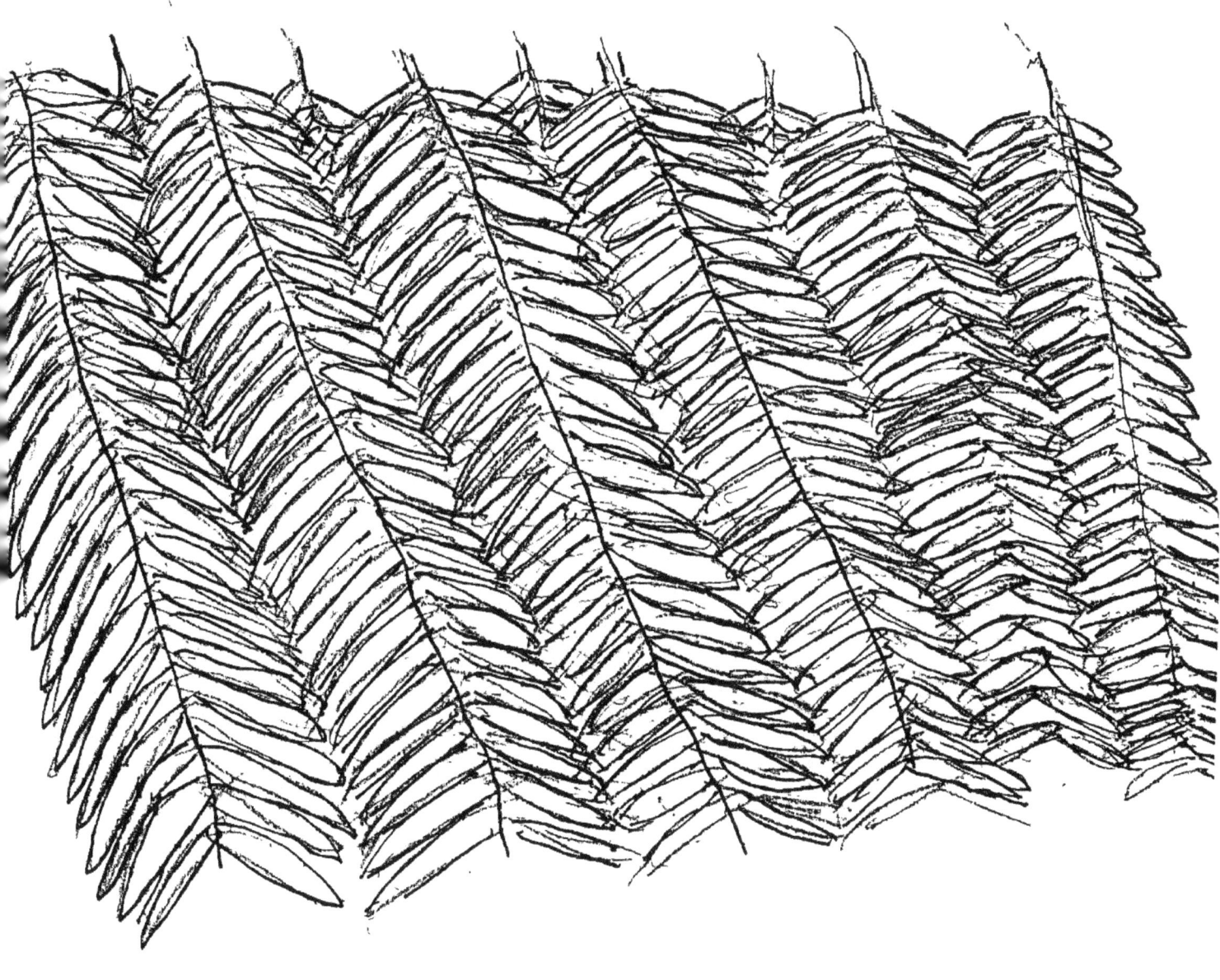

There are animals who use plant leaves for...

... shelter

Some bat species make their homes under big leaves in tropical forests.

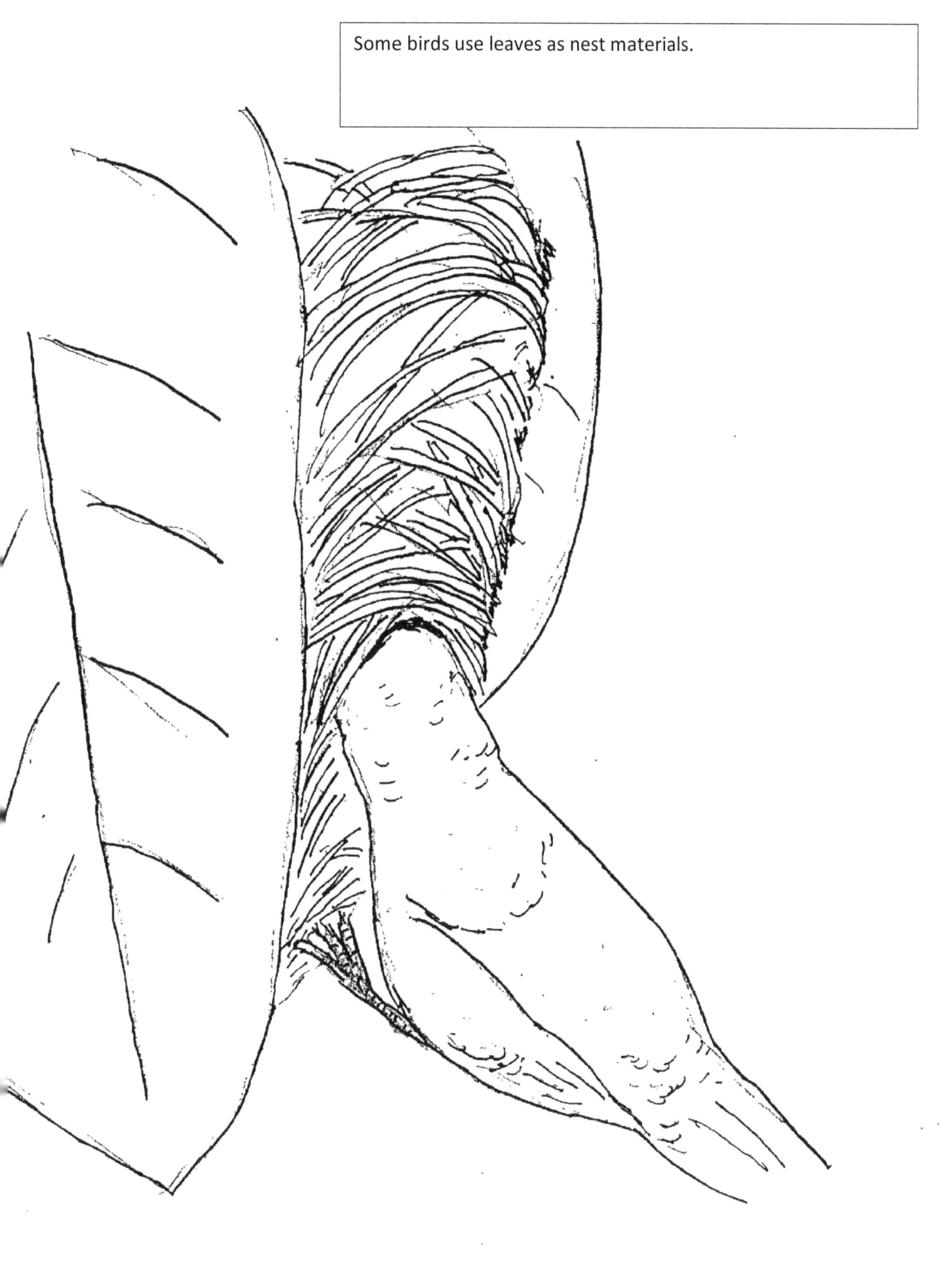

Some birds use leaves as nest materials.

...and food stuff.

Some leaf-cutter ant species use pieces of leaves to cultivate fungus for their food.

We can learn more about plant leaves through the computer, books, some teachers, and museums. However, we turn to God to learn and understand more about them.

References for Use

Burnie, David. Eyewitness Books: Plant. 1st Ed. Alfred A. Knopf; New York. 1989.

Erik von Wyk, Ben & Michael Wink. Medicinal Plants of the World: An Illustrated Scientific Guide to Important Medicinal Plants and their Uses. Timber Press; Portland. 2004.

Garfield Park Conservatory, Chicago IL

Home Remedies & Best Natural Treatment: Gurmar Medical Plant Uses and Images.
http://www.homeremediess.com/gurmar-medicinal-plant-uses-and-images/

Souza. D.M. Meat-Eating Plants. Franklin Watts: New York. 2002.

www.ingramcontent.com/pod-product-compliance
Lightning Source LLC
Chambersburg PA
CBHW062235220526
45471CB00009B/3493